Renan Bazzo

Management of solid construction waste

Renan Bazzo

Management of solid construction waste

Sustainability

ScienciaScripts

Imprint

Cover image: Provided by the author

This book is a translation from the original published under ISBN 978-613-9-71920-4.

Publisher:
Sciencia Scripts
is a trademark of
Dodo Books Indian Ocean Ltd. and OmniScriptum S.R.L publishing group

120 High Road, East Finchley, London, N2 9ED, United Kingdom
Str. Armeneasca 28/1, office 1, Chisinau MD-2012, Republic of Moldova, Europe
Printed at: see last page
ISBN: 978-620-7-72591-5

1

Thank you

I would firstly like to thank God for giving me the health and strength to overcome difficulties. To this college, its teaching staff, management and administration. To my supervisor Alexandre Teixeira de Souza, for his support, corrections and encouragement. To my father Carlos and my mother Nilza, for their love, encouragement and unconditional support. And to everyone who directly or indirectly played a part in my education, thank you very much.

Caio Correa Brito

Thank you

I would firstly like to thank God, who has illuminated my path throughout this journey. To my parents for their love, encouragement and unconditional support. To my supervisor Professor Alexandre Teixeira de Souza for his guidance and corrections. My thanks to the friends who have been part of my education and who will certainly continue to be present in my life. I would like to thank all the teachers for giving me not only rational knowledge, but also the manifestation of the character and affection of education in the process of professional training, thank you very much.

Renan Bazzo

SUMMARY

Construction waste accounts for approximately 45 per cent of the solid waste produced daily in the municipality of Tupã-SP. This is made up of stones, sand, concrete, wood, ceramics, bricks, tiles, paint, i.e. materials from construction, remodelling and demolition. The irregular disposal of RCC in Brazilian municipalities causes serious environmental impacts, so there is a need to study ways of minimising these impacts caused by incorrect disposal. With this in mind, this study aims to carry out a bibliographical survey on the management of construction waste in the municipality of Tupã-SP. The archives of the municipal administration bodies that manage construction and demolition waste (CDW) were consulted, the relevant municipal legislation was researched, as well as how this waste is managed, State Law 12.300/2006, which establishes the State Solid Waste Policy and CONAMA Resolution 307/2002, as well as the links in the production chain and the final destination of this waste.

KEY WORDS: Management, Recycling, Construction Waste.

SUMMARY

1 INTRODUCTION

Since the 18th century, with the increase in the world's population combined with the migration of the rural population to large urban centres, there has been an extraordinary increase in the amount of construction waste generated in public and private works, compromising the availability of natural resources and making it difficult to dispose of the waste generated. The great difficulty would be to reconcile this excellent productive activity with conditions that lead to the sustainable development of the planet.

One of the characteristics of civil construction activities is the consumption of materials and the generation of "pulverised" waste in various parts of the city, which makes it difficult to manage RCCs. Another difficulty is the informality of most construction work. Approximately 75 per cent of the waste generated by this activity comes from informal events (construction, renovation and demolition work, usually carried out by the owners themselves) (Mansor, M. et al, 2013, P.58).

From small renovations to large construction projects, each Brazilian produces approximately 0.5 tonnes of construction waste every year in Brazil (INAC - Instituto Nova Ágora de Cidadania).

The correct management of RCCs is supported at federal level by environmental legislation, through the National Solid Waste Policy, the State Plan, the Solid Waste Management Plan and CONAMA Resolution 307 of 2002, amended by Resolutions 448/12, 431/11 and 348/04, which set out the guidelines, criteria and procedures for the management of RCCs, regulating the actions needed to minimise environmental impacts. State and municipal legislation must also be observed (Mansor, M. et al, 2013, P.58).

Construction waste occupies a large volume for final disposal, considering that only 6 per cent of Brazilian cities surveyed in the sanitation census have some kind of waste processing facility. Waste treatment methods should be proposed and implemented (IBGE, 2000).

The recycling of solid construction waste is insignificant in Brazil compared to European countries, due to the scarcity of natural resources in their lands.

The environmental impact caused by the inadequate management of this waste is dramatically reflected, particularly in large urban centres, in flooding, the degradation of the urban landscape with severe social impacts and material losses.

The aim of this work is to evaluate the management of solid construction waste in the municipality of Tupã, seeking to present solutions that promote proper management and, above all, provide a healthy environment for the population.

2 BACKGROUND

The justification for this work is based on the large generation of solid waste from the construction industry, given that the sector currently plays an important role in economic and social terms for the country. As this area is one of the biggest generators of waste, the environmental impacts caused by the incorrect disposal of these materials are becoming increasingly worrying.

Irregular deposits of these materials cause serious problems for the population, as they provide a breeding ground for disease vectors, clog up the "manholes" responsible for urban drainage, silt up rivers and contribute directly to flooding in urban centres.

Due to the high concentration of solid construction waste produced in urban areas, there is a clear need to recycle and reuse these materials in various areas that will favour the sustainability of the planet.

The aim of this research is to reuse the materials acquired in the sorting process for the recovery of rural roads, erosion control, landscaping and the manufacture of blocks and bricks,

As well as minimising scarcity and reducing the environmental impact of exploiting natural resources, recycling solid construction waste will bring social, political and environmental advantages to municipalities. For town halls, it means fewer expenses and less land used to dispose of this waste.

3 OBJECTIVES

3.1 General Objective

The aim of this work is to analyse the management of solid construction waste in the municipality of Tupã-SP. To provide information on the transport, sorting, storage and reuse of solid construction waste in the municipality of Tupã, seeking to minimise the environmental impacts caused by the incorrect disposal of these materials.

3.2 Specific Objective

Contribute to minimising waste in the construction industry by presenting methods used to reuse the materials obtained from sorting solid construction waste.

4 LITERATURE REVIEW

4.1 General data of the municipality of Tupã

Tupã is one of the 29 (twenty-nine) municipalities in São Paulo considered a tourist resort by the state of São Paulo, as they fulfil certain prerequisites defined by state law (Câmara Municipal da Estância Turística de Tupã, 2014).

The tourist resort of Tupã has a population of 63,473 inhabitants and is located in homogeneous micro-region 621, called Alta Paulista. Its territorial area is 628.513 km^2 , but it has an urban area of 32.27 km^2 , and is located at an altitude of 511 metres (IBGE).

Tupã is located at latitude 21°56'01" south and longitude W.Gr. 50°30'45". Its territory is predominantly sandy (Bauru sandstone), with hydrography formed by the Rio do Peixe and Aguapeí rivers, as well as the Lacri, 7 de Setembro, Pitangueiras and Afonso XIII streams (Câmara Municipal da Estância Turística de Tupã, 2014).

Access to the municipality is facilitated by the paved highways SP-294 (Rodovia Comandante João Ribeiro de Barros), SPV-52, the road linking Tupã to Quatá-SP, and the side roads Tupã a Arco Íris and Tupã a Juliânia.

4.2 Data on solid waste generation and collection in the municipality of Tupã-SP

4.2.1 Organic waste

This is waste of animal or vegetable origin that was recently used by living beings. This includes food waste, coffee grounds, leaves, wood, meat and so on. These materials require special care when disposed of, as they are conducive to the development of disease-causing organisms, as well as providing a bad odour due to their rapid decomposition (SEMAMA, 2014).[1]

In the municipality of Tupã, 240,000kg of organic waste is collected every week and sent to the landfill in Tupã-SP.

4.2.2 Recyclable waste

Recyclable waste is material that, after undergoing a physical or chemical transformation, can be used again on the market, either in its original form or as raw material for new objects. The most common materials that are recycled are paper, glass, plastic and metals (SEMAMA, 2014).[2]

An average of 132,000kg of this waste is collected every week in Tupã-SP.

4.2.3 Construction Waste

This is waste from construction or demolition work, whether public or private. If not treated correctly, these materials pollute rivers and springs, favour the reproduction of insects, venomous animals and disease-transmitting microorganisms, and clog up the urban drainage system, causing flooding. Examples of these materials are: blocks, bricks, cement, concrete and others. This waste is sent by skip hire companies to the Tupã Construction Solid Waste Sorting Plant (SEMAMA, 2014). [3]

On average, 412,000kg of construction waste is collected every week.

4.2.4 Waste from Pruning/Cutting Trees, Weeding and Wood Waste

This is waste from pruning or extracting trees in public or private areas of the municipality, which is collected daily by an employee of the Tupã tourist resort. Weeding waste is the waste resulting from weeding public roads and streets, mowing, scraping and the remains of cleaning services in squares, parks and gardens (SEMAMA, 2014).[4]

In the municipality of Tupã, an average of 96,000 kg of these materials are collected every week.

---

[2] Data collected by the authors.
[3] Data collected by the authors.
[4] Data collected by the authors.

4.2.5 Electronic Waste

Electronic waste is the term applied to rubbish generated by electrical and electronic equipment. This waste is very harmful to the environment because many parts of these components contain heavy metals that are highly toxic to nature. Examples of these materials are: light bulbs, batteries, mobile phones, televisions, computers, among others. Once collected, these materials are sent to a company that specialises in recycling this waste (SEMAMA, 2014).[5]

On average, 185kg of these materials are collected every week by the Tupã-SP Municipal Department of Agriculture and the Environment.

4.2.6 Hospital Waste

Hospital waste is the result of medical activities carried out in healthcare facilities. The biggest problem with so-called "infectious waste" is the high risk of contaminating the population and polluting the environment. Syringes, needles and expired medicines are examples of such waste. Once collected, these materials are sent to a company specialising in the correct disposal of this waste (SEMAMA, 2014).[6]

On average, 44kg of hospital waste is collected every week in the municipality of Tupã.

4.2.7 Pneumatic waste

Pneumatic waste comes from tyres, most of which have a structure made up of different materials such as rubber, steel, nylon or polyester. Their disposal is harmful to the environment and to people's health, as they store water and serve as breeding grounds for various insects. The Municipal Department of Agriculture and the Environment collects these materials on a daily basis, and most of the time they are discarded by tyre collectors (SEMAMA, 2014).[7]

---

[5] Data collected by the authors.
[6] Data collected by the authors.
[7] Data collected by the authors.

Every week SEMAMA collects an average of 278 kg of these materials, which are then sent to a company that specialises in recycling them.

4.3 Data on waste disposal sites in the municipality of Tupã-SP (SEMAMA, 2014)[8]

- Tupã-SP Landfill: a duly licensed site designed to receive organic waste, located on Estrada Vicinal São Gonçalo s/n°;
- Cooperativa de Reciclagem de Tupã-SP: Licensed site for receiving recyclable waste, where the material is then sorted and sold. Located on Estrada Vicinal São Gonçalo, s/n°, next to the landfill site;
- Solid Construction Waste Sorting Plant: A licensed site designed to receive solid construction waste. After arriving, the materials are sorted and then recycled. Located on Estrada Vicinal São Gonçalo s/n°, next to the landfill site;
- "Aterro do Piva" (Piva Landfill): This is a duly licensed site that receives waste from pruning, tree extraction, weeding and wood scraps. Located on Estrada do Picadão s/n°, opposite the Exapit site;
- Electronic waste: SEMAMA employees collect this waste every week from different parts of the city and store it correctly at SEMAMA's headquarters, where after a considerable amount, it is sent to a company that specialises in recycling this waste;
- Hospital waste: Health centres, pharmacies and municipal hospitals have their own bins for collecting this waste. After a considerable amount has been collected, SEMAMA officials ask a specialised company to collect the waste and dispose of it correctly;
- Eco Ponto: A place where tyre waste is collected and sent to a company specialising in tyre recycling. It is located at SEMAMA's headquarters in Rua Francisco Budaibes 101, Bairro Tupã Mirim I.

4.4 Classification of Construction Waste (CCW)

Article 3 of CONAMA Resolution 307/2002 classifies construction waste into four classes, making it easier to separate waste according to its intended destination:

4.4.1 Class A

This is waste that can be reused or recycled as aggregates, such as from construction, demolition, renovation and repair of paving and other infrastructure works, including earthworks; from buildings with ceramic components (bricks, blocks, tiles, etc.), mortar and concrete; from the manufacturing and/or demolition process of pre-moulded concrete parts produced on construction sites (Mansor, M. et al, 2013, P.60);

4.4.2 Class B

This is waste that can be recycled for other uses, such as plastic, paper and cardboard, metals, glass, wood and others (Mansor, M. et al, 2013, P.60);

4.4.3 Class C

Waste for which technologies or economically viable applications for recycling have not been developed, such as the remains of products made with gypsum, which must be stored, transported and given an appropriate final destination in accordance with specific technical standards (Mansor, M. et al, 2013, P.60);

4.4.4 Class D

Hazardous waste from construction, such as paints, solvents, oils and others, or those actually or potentially contaminated, from demolitions, renovations and repairs in radiological clinics, industrial facilities and others, as well as roof tiles and other objects and materials containing asbestos or other products harmful to health, which must be stored, transported and disposed of in accordance with specific technical standards (Mansor, M. et al, 2013, P.60).

4.5 Classification of the Materials Obtained after Sorting the RCCs

According to CONAMA Resolution 307/2002, after sorting, waste can be categorised as follows:

4.5.1 Class A

Be reused or recycled in the form of aggregates or sent to a class A waste landfill, storing material for future use (Mansor, M. et al, 2013, P.61);

4.5.2 Class B

Be reused, recycled or sent to temporary storage areas, being disposed of in such a way that they can be used or recycled in the future (Mansor, M. et al, 2013, P.61).

Therefore, these two classes can be recycled and should be destined for such purposes, with class B waste being sorted, and class A waste being recycled in construction waste recycling plants.

4.6 Definitions According to CONAMA Resolution 307/2002

4.6.1 Construction waste

These are materials from construction, renovation, repair and demolition work, and those resulting from the preparation and excavation of land, commonly referred to as construction rubble, limestone or shale.

4.6.2 Generators

These are natural or legal persons, owners of public or private works, responsible for activities or undertakings that generate waste.

4.6.3 Conveyors

These are natural or legal persons in charge of collecting and transporting waste between the generating sources and the correct disposal areas.

4.6.4 Recycled aggregate

This is granular material from the processing of construction waste that has technical characteristics for use in building works, infrastructure, landfills and other engineering works;

4.6.5 Waste management

It is the management system that aims to reduce, reuse or recycle materials, including planning, responsibilities, practices, procedures and resources to develop and schedule the actions necessary to carry out the steps set out in programmes and plans;

4.6.6 Re-use

It's the process of reapplying the material, without the need for transformation;

4.6.7 Recycling

This is the process of reusing a material after it has undergone a transformation.

4.6.8 Processing

It is the act of subjecting waste to operations and/or processes that aim to provide it with conditions that allow it to be used as raw material or other products.

4.6.9 Construction waste landfill

These are the areas where techniques will be used to dispose of class "A" construction waste in the ground, with the aim of reserving segregated materials so that they can be used in the future or in the future use of the area, using engineering principles to confine them without causing damage to public health and the environment.

4.6.10 Waste disposal area

These are areas intended for the processing or final disposal of waste.

5 CASE STUDY 5.1 Construction Waste Recycling Plant in Tupã-SP

Inaugurated on 18 May 2012 by State Environment Secretary Bruno Covas, the Construction Solid Waste Recycling Plant is located on Estrada Vicinal São Gonçalo s/n°, next to the Tupã-SP landfill, and has a total area of 11,000 m^2 (SEMAMA, 2014).[9]

The Civil Construction Waste Recycling Plant in Tupã-SP complies with CONAMA Resolution 307 of 5th July 2002 and has an operating licence issued by CETESB.

The project was built entirely with funds from the Tupã City Council, and its machinery and equipment were purchased from the State Fund for Pollution Prevention and Control - FECOP (SEMAMA, 2014).[10]

The plant in Tupã-SP has the capacity to grind 10 tonnes/hour of construction waste.

Its operation starts when private skip companies dump their waste at the site. From that moment on, SEMAMA employees first sort the waste, then with the help of a wheel loader, the already separated materials are inserted into the vibrating feeder, starting the process of recycling construction waste (SEMAMA, 2014). [VII VIII]

At the moment, six employees work at the plant every day: one is a wheel loader operator, three are responsible for sorting materials and two are responsible for recycling construction waste.

Figure 1 shows the Solid Construction Waste Recycling Plant in the municipality of Tupã-SP, located on Estrada Vicinal São Gonçalo S/N.

[9] Data collected by the authors.

[10] Data collected by the authors.

[VIII] Data collected by the authors.

Figure 1 - Construction Waste Recycling Plant in Tupã-SP Source: Author.

5.2 Flowchart of the Construction Waste Recycling Plant in Tupã-SP

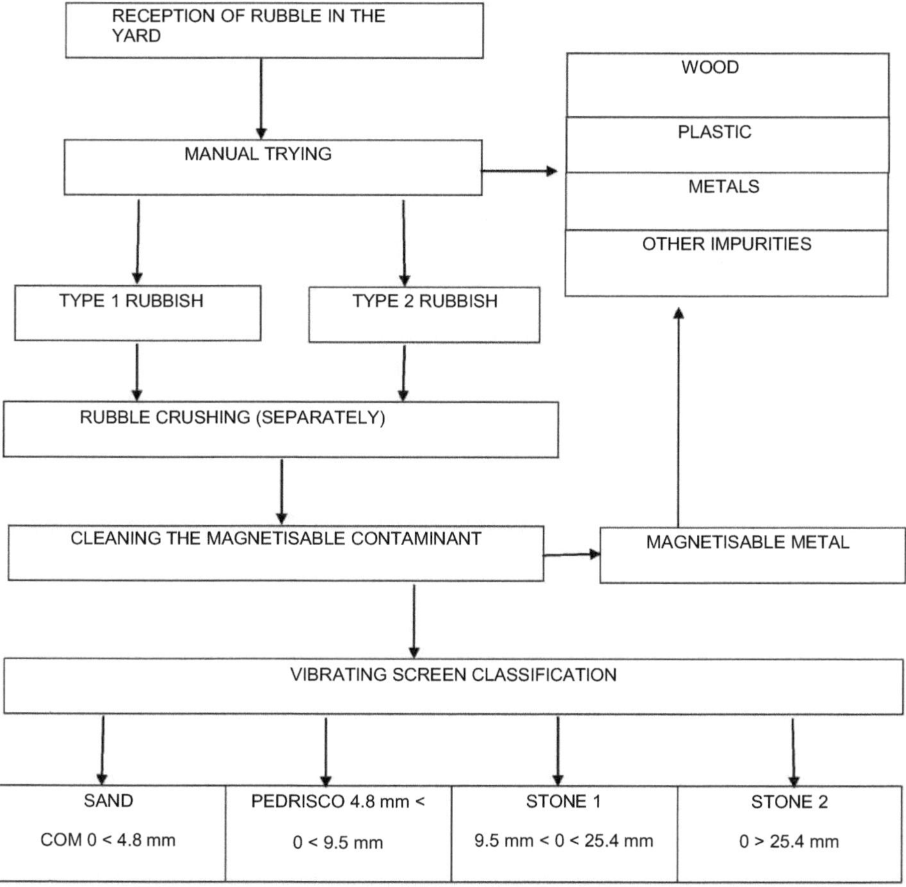

5.3 Materials obtained from the recycling of RCCs

Among the various types of waste generated by human beings, those from construction and demolition activities stand out, causing serious socio-environmental impacts. These materials require the search for quick and effective solutions for their proper management, which better focus on their final disposal and reuse in the construction production chain. Recycling CDW can produce four different types of waste, which are:

- SAND: Hard and rough particle material, with a maximum characteristic size of less than 4.8 mm, free of impurities, from the recycling of concrete, concrete blocks, bricks, tiles, etc;
- PEDRISCO: Hard and rough particle material, between 4.8 mm and 9.5 mm in size, free of impurities, from the recycling of concrete artefacts, sewage pipes and roof tiles, among others;
- STONE 1: Hard and rough particle material, between 9.5 mm and 25.4 mm in size, totally free of impurities, from the recycling of concrete, concrete blocks, bricks, roof tiles, among others;
- RACHÃO: Hard and rough particle material, larger than 25.4 mm, free of impurities and coming from the recycling of various solid construction materials.

Figures 2, 3, 4 and 5 show the four aggregates obtained by recycling construction waste.

Figure 2 - Sand obtained from the recycling of RCC. Source: Author.

Figure 3 - Gravel obtained from the recycling of RCC. Source: Author.

Figure 4 - Stone 1, obtained from the recycling of RCC. Source: Author.

Figure 5 - Rachão, material obtained from the recycling of RCC. Source: Author.

5.4 Ways of reusing RCCs

- Paving or rehabilitating rural roads: the simplest way to recycle rubble is to use it for paving or rehabilitating rural roads, in the form of stone 1 or in mixtures of recycled aggregate with soil (Abrecon, 2014);
- Use as aggregate for concrete: both the sand and gravel obtained in the recycling plant process can be used as aggregate for non-structural concrete, replacing conventional aggregates (sand and gravel) (Abrecon, 2014);
- Landscaping: crushed stone, a material obtained from recycling RCCs, can be used to beautify flowerbeds and gardens, improving water infiltration into the soil (Abrecon, 2014);
- Filling voids in buildings: Crack, being the thickest material obtained from the recycling of MSCS, can be reused to fill voids in buildings, minimising expenditure on other materials (Abrecon, 2014);
- Topographic land preparation: Both stone 1 and gravel can be reused for land preparation due to their greater thickness and ease of soil drainage (Abrecon, 2014);
- Car parks: Stone 1 can replace gravel, which is normally used for soil stabilisation in car parks, reducing the risk of flooding due to its ease of infiltration (Abrecon, 2014).

5.5 Examples of CCW reuse in Tupã-SP

Figures 6, 7, 8, 9 and 10 show some of the methods of reusing aggregates acquired through the recycling of construction waste in the municipality of Tupã-SP.

Figure 6 - Landscaping at Guatemala Street Square in Tupã-SP. Source: Author.

Figure 7 - Recovery of the "Sabiá Rural Road" in Tupã-SP. Source: Author.

Figure 8 - Car park of the Municipal Department of Agriculture and the Environment in Tupã-SP. Source: Author.

Figure 9 - Santa Adélia Health Centre car park in Tupã-SP. Source: Author.

Figure 10 - Landscaping at Francisco Cardoso Square in the municipality of Tupã-SP Source: Author's own information.

5.6 Environmental benefits of recycling RCCs

In the current production model, waste is always generated, whether for durable consumer goods (buildings, bridges and roads) or non-durable goods (disposable packaging). In this process, production almost always uses non-renewable raw materials of natural origin. This model did not present problems until recently, due to the large amount of natural resources available and the smaller number of people incorporated into consumer society (JOHN, 1999; JOHN, 2000; CURWELL; COOPER, 1998; GÜNTHER, 2000).

With growing industrialisation, along with new technologies, population growth, an increase in the number of people in urban centres and the diversification of consumption of goods and services, construction waste has become a serious urban problem with costly and complex management, considering the volume and mass accumulated, especially after 1980. The problems were characterised by a shortage of waste disposal areas caused by the occupation and development of urban areas, high social costs in waste management, public sanitation problems and environmental contamination (JOHN, 1999; JOHN, 2000; PINTO, 1999).

During ECO-92 and the definition of Agenda 21, the urgent need to programme an appropriate environmental management system for solid waste was highlighted. One of the solutions to the problems generated is waste recycling, in which the construction industry has great potential for utilising waste, since it consumes up to 75% of natural resources (JOHN, 2000; GÜNTHER, 2000; PINTO, 1999).

- Reduction in the consumption of non-renewable natural resources when replaced by recycled waste (JOHN, 2000).
- Reducing the areas needed for landfill by minimising the volume of waste through recycling. Of particular note here is the need to recycle construction and demolition waste itself, which accounts for more than 50 per cent of the mass of municipal solid waste (PINTO, 1999).
- Reduced energy consumption during the production process. In particular, the cement industry uses waste with good calorific value to obtain its raw material (co-incineration) or by using blast furnace slag, waste with a composition similar to cement (JOHN, 2000).

- Reduction of pollution for cement industries, as it reduces carbon dioxide emissions by using blast furnace slag to replace portland cement (JOHN, 1999).

In fact, it is known that isolated actions will not solve the problems caused by this waste and that the industry must try to close its production cycle in such a way as to minimise the output of waste and the input of non-renewable raw materials. In general, these cycles for construction try to bring construction closer to the concept of sustainable development, understood here as a process that leads to changes in the exploitation of resources, the direction of investments, the orientation of technological development and institutional changes, all aimed at harmony and intertwining in present and future human aspirations and needs. This concept involves cultural changes, environmental education and a systemic vision (ANGULO, 2000; JOHN, 2000).

Although reducing waste generation is always a necessary action, it is limited since there are impurities in the raw material, it involves costs and levels of technological development (JOHN, 2000).

In this way, recycling construction waste can generate the numerous benefits listed below:

5.7 Public Policies on Construction Waste

5.7.1 Municipal Law

Law No. 4.689 of 15 April 2014, which prohibits the dumping or deposit of rubbish or debris on side roads, municipal roads and vacant lots in the municipality, and makes other provisions (Tupã, 2014).

The purpose of Law No. 4,489/2014 is to prohibit the dumping or depositing, by any means or for any reason, of rubbish or debris on public or private roads in the municipality in order to preserve the environment and public hygiene (Tupã, 2014).

According to Article 2, transgressions of the rules set out in this Law generate the following penalties for the offender (Tupã, 2014):

- Notification to remove the irregularly deposited waste within 24 (twenty-four) hours;
- A fine of 15 (fifteen) UFM (1 UFM = R$ 60.95);

- Twenty-four (24) hours after the fine has been issued, if the offence persists, the town hall will remove the waste and charge a fee for the service provided.

5.7.2 State Law

Created in 2006, with a broad set of principles, guidelines and instruments for solid waste management, established by State Law 12.300, of 16 March 2006, which institutes the State Solid Waste Policy - PERS, regulated by Decree No. 54.645, of 5 August 2009 (São Paulo, 2006).

The aim of the State Solid Waste Policy is to reduce the quantity and harmfulness of solid waste, avoid the environmental and public health problems they generate and eradicate "dumps", "controlled landfills", "dumpsites" and other inappropriate destinations, as well as the sustainable, rational and efficient use of natural resources (São Paulo, 2006).

The State Solid Waste Policy defines solid construction waste as that which comes from the construction, remodelling, repair and demolition of civil construction works, and that which results from the preparation and excavation of land, such as: bricks, ceramic blocks, concrete in general, soils, rocks, metals, resins, glues, paints, wood, plywood, linings and mortars, plaster, roof tiles, asphalt pavement, glass, plastics, pipes and electrical wiring, commonly called construction rubble, caliça or metralha (São Paulo, 2006).

PERS defines fundamental planning instruments for structuring solid waste management, such as Solid Waste Plans, the Annual Solid Waste Declaration System, the State Solid Waste Inventory and the monitoring of environmental quality indicators. These are the instruments that will support the development of public policies that will promote the minimisation of waste generated, i.e. the reduction of the quantity and dangerousness of materials and substances discarded in the environment (Mansor, M. et al, 2013, P.16).

5.7.3 Federal Law

CONAMA Resolution No. 307 of 5 July 2002 establishes guidelines, criteria and procedures for the management of construction waste, regulating the actions required to minimise environmental impacts.

This Resolution arose from the great need to solve problems arising from the immense generation of waste and its environmental, social and economic impacts. The lack of management must be replaced by an integrated municipal management programme (Brasil, 2002).

Considerations in accordance with CONAMA Resolution No. 307 of 5 July 2002:

- Urban policy for the full development of the social function of the city and urban property, as set out in Law No. 10.257 of 10 July 2001 (Dias, 2007);
- The need to programme guidelines to effectively reduce the environmental impacts

generated by waste (Dias, 2007);

- Disposing of waste in inappropriate places contributes to the degradation of environmental quality (Dias, 2007);
- CDW represents a significant percentage of the solid waste produced in urban areas (Dias, 2007);
- Generators of CDW must be responsible for waste from construction, renovation, repair and demolition of structures and roads, as well as waste resulting from vegetation removal and soil excavation (Dias, 2007);
- Technical and economic feasibility of producing and using recycled materials from the construction industry (Dias, 2007);
- Integrated waste management should provide social, economic and environmental benefits (Dias, 2007).

Article 5 of CONAMA Resolution 307 establishes that, in order to manage waste, an integrated construction waste management plan will be implemented, to be drawn up by the municipalities and the Federal District. This plan will include a municipal waste management programme and waste management projects by generators (Brasil, 2002).

According to Article 9, Construction Waste Management Projects must include the following stages (Brasil, 2002):

- Characterisation: at this stage the generator must identify and quantify the waste;
- Sorting: This should preferably be carried out by the generator at source, or be carried out in licensed disposal areas for this purpose, respecting the classes of waste established in Article 3 of this Resolution;
- Packaging: The generator must ensure that the waste is contained after generation until the transport stage, ensuring in all cases where possible, the conditions for reuse and recycling;
- Transport: This must be carried out in accordance with the previous steps and in compliance with the technical regulations in force for transporting waste;
- Disposal: Must be provided for in accordance with the provisions of this resolution;

6 CONCLUSION

The development of this research on the management of construction waste in the city of Tupã, state of São Paulo, was based on the bibliographical review of this work, through the application of the concepts and data collected.

Using the chosen working method, it was possible to achieve the main objectives of this end-of-course work, both in presenting and analysing the legislation presented and in monitoring the Construction Waste Recycling Plant in Tupã-SP.

A Construction Waste Management Project contributes directly to minimising environmental impacts, both in terms of waste during the construction phase and when these materials are disposed of.

Construction companies are seen as the main culprits in generating large amounts of rubble, and sometimes, in order to reduce disposal costs, they transfer all responsibility to society. Sustainability in construction must be taken into account from the start of a project right through to planning the population's quality of life. To this end, it is essential to use sustainable resources when building new projects.

As a result of the large amount of waste found on rural roads and wastelands, the Tupã Tourist Resort City Council enacted Law No. 4,689 of 15 April 2014, which prohibits the dumping or depositing of rubbish and debris at local and municipal entrances and wastelands in the municipality in order to preserve the environment and public hygiene. (Câmara Municipal de Estância Turística de Tupã, 2014).

It is very important to understand these laws and consequently to comply with them, because one of the biggest difficulties SEMAMA employees encounter is the irregular disposal of this waste.

In view of all the research that has gone into completing this work, it can be said that managing construction waste is of great benefit, as the positive aspects are many and can be listed as: reducing collection costs, reducing waste with less waste generation, reusing waste, and cleaning and organising roadbeds and vacant lots.

In a global and simplified way, the main justification for choosing the topic presented in this work is the conservation of a pleasant environment, always seeking a healthy quality of life for the population.

BIBLIOGRAPHY

ABRECON - **Associação Brasileira para Reciclagem para Resíduos da Construção Civil e Demolição**, available at: <http://www.abrecon.org.br/Conteudo/8/Aplicacao.aspx>, Accessed on 22 September 2014, 09h45m20s).

ANGULO, S.C. **Variability of recycled construction and demolition waste aggregates**. São Paulo, 2000. 155p. Master's dissertation - Polytechnic School, University of São Paulo.

DIAS, E.C.M, **Gestão de Resíduos na construção civil,** São Paulo, 2007, p 21-23. Graduation - Anhembi Murumbi University.

GUNTHER, W.M.R. **Waste minimisation and environmental education.** In: NATIONAL SEMINAR ON SOLID WASTE AND PUBLIC CLEANING, 7. Curitiba, 2000. **Proceedings**. Curitiba, 2000.

JOHN, V.M.J. Overview of waste recycling in the construction industry. In: SEMINAR ON SUSTAINABLE DEVELOPMENT AND RECYCLING IN CIVIL CONSTRUCTION, 2, São Paulo, 1999. **Proceedings**. São Paulo, IBRACON, 1999. p.44-55.

IBGE - Brazilian Institute of Geography and Statistics, available at: <http://www.ibge.gov.br/home/estatistica/populacao/condicaodevida/pnsb2008/tabelas pdf/tab099.pdF>, Accessed on 12 Aug. 2014, 13h05m10s;

INAC - **Instituto Nova Ágora de Cidadania**, available at: http://inac.org.br/site/, Accessed on 12 Aug. 2014, 10h25m25s;

JOHN, V.M. **Reciclagem de resíduos na construção civil - contribuição à metodologia de pesquisa e desenvolvimento**. São Paulo, 2000. 102p. Thesis (livre docência) - Polytechnic School, University of São Paulo.

LeiM unicipalN °4 .689/2014, available at : http://www.camaratupa.sp.gov.br/camver/leimun/2014/04689.pdf , Accessed on 228 Sep. 2014, 07h40m10s.

State Law No. 12 .300, Available at : <http://governo.sp.jusbrasil.com.br/legislacao/135418/politica-estadual-de-residuos-solidos- lei-12300-06>. Accessed on 17 September 2014, 14:25:28.

MANSOR, M. et al, **Caderno de Educação Ambiental - Resíduos Sólidos.** 2° edição,São Paulo, 2013, p 58-63.

PINTO, T.P. **Metodologia para a gestão diferenciada de resíduos sólidos da construção urbana**. São Paulo, 1999. 189p. Thesis (Doctorate) - Polytechnic School, University of São Paulo.

CONAMA Resolution No. 307, Available at: <http://www.mma.gov.br/port/conama/res/res02/res30702.html>. Accessed on: 17 September 2014, 16h10m55s.

Printed by Books on Demand GmbH, Norderstedt / Germany